THE TWO DREAMS
BY
ALGERNON CHARLES SWINBURNE

dition et mise en page : Culturea (Hérault, 34)
etrouvez notre catalogue sur http://culturea.fr/
ontact : infos@culturea.fr
nprimé en Allemagne par Books on Demand Gmbh
esign typographique et layout : Derek Murphy et Reedsy
SBN : 9791041986200
ate de parution : février 2024

The Two Dreams by Algernon Charles Swinburne

THE TWO DREAMS
(FROM BOCCACCIO)

I will that if I say a heavy thing
Your tongues forgive me; seeing ye know that spring
Has flecks and fits of pain to keep her sweet,
And walks somewhile with winter-bitten feet.
Moreover it sounds often well to let
One string, when ye play music, keep at fret
The whole song through; one petal that is dead
Confirms the roses, be they white or red;
Dead sorrow is not sorrowful to hear
As the thick noise that breaks mid weeping were;
The sick sound aching in a lifted throat
Turns to sharp silver of a perfect note;
And though the rain falls often, and with rain
Late autumn falls on the old red leaves like pain,
I deem that God is not disquieted.
Also while men are fed with wine and bread,
They shall be fed with sorrow at his hand.
There grew a rose-garden in Florence land
More fair than many; all red summers through
The leaves smelt sweet and sharp of rain, and blew
Sideways with tender wind; and therein fell
Sweet sound wherewith the green waxed audible,
As a bird's will to sing disturbed his throat
And set the sharp wings forward like a boat
Pushed through soft water, moving his brown side
Smooth-shapen as a maid's, and shook with pride
His deep warm bosom, till the heavy sun's
Set face of heat stopped all the songs at once.
The ways were clean to walk and delicate;
And when the windy white of March grew late,
Before the trees took heart to face the sun
With ravelled raiment of lean winter on,
The roots were thick and hot with hollow grass.
Some roods away a lordly house there was,
Cool with broad courts and latticed passage wet
From rush-flowers and lilies ripe to set,

Sown close among the strewings of the floor;
And either wall of the slow corridor
Was dim with deep device of gracious things;
Some angel's steady mouth and weight of wings
Shut to the side; or Peter with straight stole
And beard cut black against the aureole
That spanned his head from nape to crown; thereby
Mary's gold hair, thick to the girdle-tie
Wherein was bound a child with tender feet;
Or the broad cross with blood nigh brown on it.
Within this house a righteous lord abode,
Ser Averardo; patient of his mood,
And just of judgment; and to child he had
A maid so sweet that her mere sight made glad
Men sorrowing, and unbound the brows of hate;
And where she came, the lips that pain made strait
Waxed warm and wide, and from untender grew
Tender as those that sleep brings patience to.
Such long locks had she, that with knee to chin
She might have wrapped and warmed her feet therein.
Right seldom fell her face on weeping wise;
Gold hair she had, and golden-coloured eyes,
Filled with clear light and fire and large repose
Like a fair hound's; no man there is but knows
Her face was white, and thereto she was tall;
In no wise lacked there any praise at all
To her most perfect and pure maidenhood;
No sin I think there was in all her blood.
She, where a gold grate shut the roses in,
Dwelt daily through deep summer weeks, through green
Flushed hours of rain upon the leaves; and there
Love made him room and space to worship her
With tender worship of bowed knees, and wrought
Such pleasure as the pained sense palates not
For weariness, but at one taste undoes
The heart of its strong sweet, is ravenous
Of all the hidden honey; words and sense
Fail through the tune's imperious prevalence.
In a poor house this lover kept apart,
Long communing with patience next his heart
If love of his might move that face at all,

Tuned evenwise with colours musical;
Then after length of days he said thus: "Love,
For love's own sake and for the love thereof
Let no harsh words untune your gracious mood;
For good it were, if anything be good,
To comfort me in this pain's plague of mine;
Seeing thus, how neither sleep nor bread nor wine
Seems pleasant to me, yea no thing that is
Seems pleasant to me; only I know this,
Love's ways are sharp for palms of piteous feet
To travel, but the end of such is sweet:
Now do with me as seemeth you the best."
She mused a little, as one holds his guest
By the hand musing, with her face borne down:
Then said: "Yea, though such bitter seed be sown,
Have no more care of all that you have said;
Since if there is no sleep will bind your head,
Lo, I am fain to help you certainly;
Christ knoweth, sir, if I would have you die;
There is no pleasure when a man is dead."
Thereat he kissed her hands and yellow head
And clipped her fair long body many times;
I have no wit to shape in written rhymes
A scanted tithe of this great joy they had.
They were too near love's secret to be glad;
As whoso deems the core will surely melt
From the warm fruit his lips caress, hath felt
Some bitter kernel where the teeth shut hard:
Or as sweet music sharpens afterward,
Being half disrelished both for sharp and sweet;
As sea-water, having killed over-heat
In a man's body, chills it with faint ache;
So their sense, burdened only for love's sake,
Failed for pure love; yet so time served their wit,
They saved each day some gold reserves of it,
Being wiser in love's riddle than such be
Whom fragments feed with his chance charity.
All things felt sweet were felt sweet overmuch;
The rose-thorn's prickle dangerous to touch,
And flecks of fire in the thin leaf-shadows;
Too keen the breathed honey of the rose,

Its red too harsh a weight on feasted eyes;
They were so far gone in love's histories,
Beyond all shape and colour and mere breath,
Where pleasure has for kinsfolk sleep and death,
And strength of soul and body waxen blind
For weariness, and flesh entailed with mind,
When the keen edge of sense foretasteth sin.
Even this green place the summer caught them in
Seemed half deflowered and sick with beaten leaves
In their strayed eyes; these gold flower-fumèd eves
Burnt out to make the sun's love-offering,
The midnoon's prayer, the rose's thanksgiving,
The trees' weight burdening the strengthless air,
The shape of her stilled eyes, her coloured hair,
Her body's balance from the moving feet—
All this, found fair, lacked yet one grain of sweet
It had some warm weeks back: so perisheth
On May's new lip the tender April breath:
So those same walks the wind sowed lilies in
All April through, and all their latter kin
Of languid leaves whereon the Autumn blows—
The dead red raiment of the last year's rose—
The last year's laurel, and the last year's love,
Fade, and grow things that death grows weary of.
What man will gather in red summer-time
The fruit of some obscure and hoary rhyme
Heard last midwinter, taste the heart in it,
Mould the smooth semitones afresh, refit
The fair limbs ruined, flush the dead blood through
With colour, make all broken beauties new
For love's new lesson—shall not such find pain
When the marred music labouring in his brain
Frets him with sweet sharp fragments, and lets slip
One word that might leave satisfied his lip—
One touch that might put fire in all the chords?
This was her pain: to miss from all sweet words
Some taste of sound, diverse and delicate—
Some speech the old love found out to compensate
For seasons of shut lips and drowsiness—
Some grace, some word the old love found out to bless
Passionless months and undelighted weeks.

The flowers had lost their summer-scented cheeks,
Their lips were no more sweet than daily breath:
The year was plagued with instances of death.
So fell it, these were sitting in cool grass
With leaves about, and many a bird there was
Where the green shadow thickliest impleached
Soft fruit and writhen spray and blossom bleached
Dry in the sun or washed with rains to white:
Her girdle was pure silk, the bosom bright
With purple as purple water and gold wrought in.
One branch had touched with dusk her lips and chin,
Made violet of the throat, abashed with shade
The breast's bright plaited work: but nothing frayed
The sun's large kiss on the luxurious hair.
Her beauty was new colour to the air
And music to the silent many birds.
Love was an-hungred for some perfect words
To praise her with; but only her low name
"Andrevuola" came thrice, and thrice put shame
In her clear cheek, so fruitful with new red
That for pure love straightway shame's self was dead.
Then with lids gathered as who late had wept
She began saying: "I have so little slept
My lids drowse now against the very sun;
Yea, the brain aching with a dream begun
Beats like a fitful blood; kiss but both brows,
And you shall pluck my thoughts grown dangerous
Almost away." He said thus, kissing them:
"O sole sweet thing that God is glad to name,
My one gold gift, if dreams be sharp and sore
Shall not the waking time increase much more
With taste and sound, sweet eyesight or sweet scent?
Has any heat too hard and insolent
Burnt bare the tender married leaves, undone
The maiden grass shut under from the sun?
Where in this world is room enough for pain?"
The feverish finger of love had touched again
Her lips with happier blood; the pain lay meek
In her fair face, nor altered lip nor cheek
With pallor or with pulse; but in her mouth
Love thirsted as a man wayfaring doth,

Making it humble as weak hunger is.
She lay close to him, bade do this and this,
Say that, sing thus: then almost weeping-ripe
Crouched, then laughed low. As one that fain would wipe
The old record out of old things done and dead,
She rose, she heaved her hands up, and waxed red
For wilful heart and blameless fear of blame;
Saying "Though my wits be weak, this is no shame
For a poor maid whom love so punisheth
With heats of hesitation and stopped breath
That with my dreams I live yet heavily
For pure sad heart and faith's humility.
Now be not wroth and I will show you this.
"Methought our lips upon their second kiss
Met in this place, and a fair day we had
And fair soft leaves that waxed and were not sad
With shaken rain or bitten through with drouth;
When I, beholding ever how your mouth
Waited for mine, the throat being fallen back,
Saw crawl thereout a live thing flaked with black
Specks of brute slime and leper-coloured scale,
A devil's hide with foul flame-writhen grail
Fashioned where hell's heat festers loathsomest;
And that brief speech may ease me of the rest,
Thus were you slain and eaten of the thing.
My waked eyes felt the new day shuddering
On their low lids, felt the whole east so beat,
Pant with close pulse of such a plague-struck heat,
As if the palpitating dawn drew breath
For horror, breathing between life and death,
Till the sun sprang blood-bright and violent."
So finishing, her soft strength wholly spent,
She gazed each way, lest some brute-hoovèd thing,
The timeless travail of hell's childbearing,
Should threat upon the sudden: whereat he,
For relish of her tasted misery
And tender little thornprick of her pain,
Laughed with mere love. What lover among men
But hath his sense fed sovereignly 'twixt whiles
With tears and covered eyelids and sick smiles
And soft disaster of a painèd face?

What pain, established in so sweet a place,
But the plucked leaf of it smells fragrantly?
What colour burning man's wide-open eye
But may be pleasurably seen? what sense
Keeps in its hot sharp extreme violence
No savour of sweet things? The bereaved blood
And emptied flesh in their most broken mood
Fail not so wholly, famish not when thus
Past honey keeps the starved lip covetous.
Therefore this speech from a glad mouth began,
Breathed in her tender hair and temples wan
Like one prolonged kiss while the lips had breath.
"Sleep, that abides in vassalage of death
And in death's service wears out half his age,
Hath his dreams full of deadly vassalage,
Shadow and sound of things ungracious;
Fair shallow faces, hooded bloodless brows,
And mouths past kissing; yea, myself have had
As harsh a dream as holds your eyelids sad.
"This dream I tell you came three nights ago;
In full mid sleep I took a whim to know
How sweet things might be; so I turned and thought;
But save my dream all sweet availed me not.
First came a smell of pounded spice and scent
Such as God ripens in some continent
Of utmost amber in the Syrian sea;
And breaths as though some costly rose could be
Spoiled slowly, wasted by some bitter fire
To burn the sweet out leaf by leaf, and tire
The flower's poor heart with heat and waste, to make
Strong magic for some perfumed woman's sake.
Then a cool naked sense beneath my feet
Of bud and blossom; and sound of veins that beat
As if a lute should play of its own heart
And fearfully, not smitten of either part;
And all my blood it filled with sharp and sweet
As gold swoln grain fills out the huskèd wheat;
So I rose naked from the bed, and stood
Counting the mobile measure in my blood
Some pleasant while, and through each limb there came
Swift little pleasures pungent as a flame,

Felt in the thrilling flesh and veins as much
As the outer curls that feel the comb's first touch
Thrill to the roots and shiver as from fire;
And blind between my dream and my desire
I seemed to stand and held my spirit still
Lest this should cease. A child whose fingers spill
Honey from cells forgotten of the bee
Is less afraid to stir the hive and see
Some wasp's bright back inside, than I to feel
Some finger-touch disturb the flesh like steel.
I prayed thus; Let me catch a secret here
So sweet, it sharpens the sweet taste of fear
And takes the mouth with edge of wine; I would
Have here some colour and smooth shape as good
As those in heaven whom the chief garden hides
With low grape-blossom veiling their white sides
And lesser tendrils that so bind and blind
Their eyes and feet, that if one come behind
To touch their hair they see not, neither fly;
This would I see in heaven and not die.
So praying, I had nigh cried out and knelt,
So wholly my prayer filled me: till I felt
In the dumb night's warm weight of glowing gloom
Somewhat that altered all my sleeping-room,
And made it like a green low place wherein
Maids mix to bathe: one sets her small warm chin
Against a ripple, that the angry pearl
May flow like flame about her: the next curl
Dips in some eddy coloured of the sun
To wash the dust well out; another one
Holds a straight ankle in her hand and swings
With lavish body sidelong, so that rings
Of sweet fierce water, swollen and splendid, fail
All round her fine and floated body pale,
Swayed flower-fashion, and her balanced side
Swerved edgeways lets the weight of water slide,
As taken in some underflow of sea
Swerves the banked gold of sea-flowers; but she
Pulls down some branch to keep her perfect head
Clear of the river: even from wall to bed,
I tell you, was my room transfigured so.

Sweet, green and warm it was, nor could one know
If there were walls or leaves, or if there was
No bed's green curtain, but mere gentle grass.
There were set also hard against the feet
Gold plates with honey and green grapes to eat,
With the cool water's noise to hear in rhymes:
And a wind warmed me full of furze and limes
And all hot sweets the heavy summer fills
To the round brim of smooth cup-shapen hills.
Next the grave walking of a woman's feet
Made my veins hesitate, and gracious heat
Made thick the lids and leaden on mine eyes:
And I thought ever, surely it were wise
Not yet to see her: this may last (who knows?)
Five minutes; the poor rose is twice a rose
Because it turns a face to her, the wind
Sings that way; hath this woman ever sinned,
I wonder? as a boy with apple-rind,
I played with pleasures, made them to my mind,
Changed each ere tasting. When she came indeed,
First her hair touched me, then I grew to feed
On the sense of her hand; her mouth at last
Touched me between the cheek and lip and past
Over my face with kisses here and there
Sown in and out across the eyes and hair.
Still I said nothing; till she set her face
More close and harder on the kissing-place,
And her mouth caught like a snake's mouth, and stung
So faint and tenderly, the fang scarce clung
More than a bird's foot: yet a wound it grew,
A great one, let this red mark witness you
Under the left breast; and the stroke thereof
So clove my sense that I woke out of love
And knew not what this dream was nor had wit;
But now God knows if I have skill of it."
Hereat she laid one palm against her lips
To stop their trembling; as when water slips
Out of a beak-mouthed vessel with faint noise
And chuckles in the narrowed throat and cloys
The carven rims with murmuring, so came
Words in her lips with no word right of them,

A beaten speech thick and disconsolate,
Till his smile ceasing waxed compassionate
Of her sore fear that grew from anything—
The sound of the strong summer thickening
In heated leaves of the smooth apple-trees:
The day's breath felt about the ash-branches,
And noises of the noon whose weight still grew
On the hot heavy-headed flowers, and drew
Their red mouths open till the rose-heart ached;
For eastward all the crowding rose was slaked
And soothed with shade: but westward all its growth
Seemed to breathe hard with heat as a man doth
Who feels his temples newly feverous.
And even with such motion in her brows
As that man hath in whom sick days begin,
She turned her throat and spake, her voice being thin
As a sick man's, sudden and tremulous;
"Sweet, if this end be come indeed on us,
Let us love more;" and held his mouth with hers.
As the first sound of flooded hill-waters
Is heard by people of the meadow-grass,
Or ever a wandering waif of ruin pass
With whirling stones and foam of the brown stream
Flaked with fierce yellow: so beholding him
She felt before tears came her eyelids wet,
Saw the face deadly thin where life was yet,
Heard his throat's harsh last moan before it clomb:
And he, with close mouth passionate and dumb,
Burned at her lips: so lay they without speech,
Each grasping other, and the eyes of each
Fed in the other's face: till suddenly
He cried out with a little broken cry
This word, "O help me, sweet, I am but dead."
And even so saying, the colour of fair red
Was gone out of his face, and his blood's beat
Fell, and stark death made sharp his upward feet
And pointed hands; and without moan he died.
Pain smote her sudden in the brows and side,
Strained her lips open and made burn her eyes:
For the pure sharpness of her miseries
She had no heart's pain, but mere body's wrack;

But at the last her beaten blood drew back
Slowly upon her face, and her stunned brows
Suddenly grown aware and piteous
Gathered themselves, her eyes shone, her hard breath
Came as though one nigh dead came back from death;
Her lips throbbed, and life trembled through her hair.
And in brief while she thought to bury there
The dead man that her love might lie with him
In a sweet bed under the rose-roots dim
And soft earth round the branchèd apple-trees,
Full of hushed heat and heavy with great ease,
And no man entering divide him thence.
Wherefore she bade one of her handmaidens
To be her help to do upon this wise.
And saying so the tears out of her eyes
Fell without noise and comforted her heart:
Yea, her great pain eased of the sorest part
Began to soften in her sense of it.
There under all the little branches sweet
The place was shapen of his burial;
They shed thereon no thing funereal,
But coloured leaves of latter rose-blossom,
Stems of soft grass, some withered red and some
Fair and fresh-blooded; and spoil splendider
Of marigold and great spent sunflower.
And afterward she came back without word
To her own house; two days went, and the third
Went, and she showed her father of this thing.
And for great grief of her soul's travailing
He gave consent she should endure in peace
Till her life's end; yea, till her time should cease,
She should abide in fellowship of pain.
And having lived a holy year or twain
She died of pure waste heart and weariness.
And for love's honour in her love's distress
This word was written over her tomb's head;
"Here dead she lieth, for whose sake Love is dead."

AHOLIBAH

In the beginning God made thee
A woman well to look upon,
Thy tender body as a tree
Whereon cool wind hath always blown
Till the clean branches be well grown.
There was none like thee in the land;
The girls that were thy bondwomen
Did bind thee with a purple band
Upon thy forehead, that all men
Should know thee for God's handmaiden.
Strange raiment clad thee like a bride,
With silk to wear on hands and feet
And plates of gold on either side:
Wine made thee glad, and thou didst eat
Honey, and choice of pleasant meat.
And fishers in the middle sea
Did get thee sea-fish and sea-weeds
In colour like the robes on thee;
And curious work of plaited reeds,
And wools wherein live purple bleeds.
And round the edges of thy cup
Men wrought thee marvels out of gold,
Strong snakes with lean throats lifted up,
Large eyes whereon the brows had hold,
And scaly things their slime kept cold.
For thee they blew soft wind in flutes
And ground sweet roots for cunning scent;
Made slow because of many lutes,
The wind among thy chambers went
Wherein no light was violent.
God called thy name Aholibah,
His tabernacle being in thee,
A witness through waste Asia;
Thou wert a tent sewn cunningly
With gold and colours of the sea.
God gave thee gracious ministers
And all their work who plait and weave:
The cunning of embroiderers
That sew the pillow to the sleeve,

And likeness of all things that live.
Thy garments upon thee were fair
With scarlet and with yellow thread;
Also the weaving of thine hair
Was as fine gold upon thy head,
And thy silk shoes were sewn with red.
All sweet things he bade sift, and ground
As a man grindeth wheat in mills
With strong wheels alway going round;
He gave thee corn, and grass that fills
The cattle on a thousand hills.
The wine of many seasons fed
Thy mouth, and made it fair and clean;
Sweet oil was poured out on thy head
And ran down like cool rain between
The strait close locks it melted in.
The strong men and the captains knew
Thy chambers wrought and fashioned
With gold and covering of blue,
And the blue raiment of thine head
Who satest on a stately bed.
All these had on their garments wrought
The shape of beasts and creeping things,
The body that availeth not,
Flat backs of worms and veinèd wings,
And the lewd bulk that sleeps and stings.
Also the chosen of the years,
The multitude being at ease,
With sackbuts and with dulcimers
And noise of shawms and psalteries
Made mirth within the ears of these.
But as a common woman doth,
Thou didst think evil and devise;
The sweet smell of thy breast and mouth
Thou madest as the harlot's wise,
And there was painting on thine eyes.
Yea, in the woven guest-chamber
And by the painted passages
Where the strange gracious paintings were,
State upon state of companies,
There came on thee the lust of these.

Because of shapes on either wall
Sea-coloured from some rare blue shell
At many a Tyrian interval,
Horsemen on horses, girdled well,
Delicate and desirable,
Thou saidest: I am sick of love:
Stay me with flagons, comfort me
With apples for my pain thereof
Till my hands gather in his tree
That fruit wherein my lips would be.
Yea, saidest thou, I will go up
When there is no more shade than one
May cover with a hollow cup,
And make my bed against the sun
Till my blood's violence be done.
Thy mouth was leant upon the wall
Against the painted mouth, thy chin
Touched the hair's painted curve and fall;
Thy deep throat, fallen lax and thin,
Worked as the blood's beat worked therein.
Therefore, O thou Aholibah,
God is not glad because of thee;
And thy fine gold shall pass away
Like those fair coins of ore that be
Washed over by the middle sea.
Then will one make thy body bare
To strip it of all gracious things,
And pluck the cover from thine hair,
And break the gift of many kings,
Thy wrist-rings and thine ankle-rings.
Likewise the man whose body joins
To thy smooth body, as was said,
Who hath a girdle on his loins
And dyed attire upon his head—
The same who, seeing, worshipped,
Because thy face was like the face
Of a clean maiden that smells sweet,
Because thy gait was as the pace
Of one that opens not her feet
And is not heard within the street—
Even he, O thou Aholibah,

Made separate from thy desire,
Shall cut thy nose and ears away
And bruise thee for thy body's hire
And burn the residue with fire.
Then shall the heathen people say,
The multitude being at ease;
Lo, this is that Aholibah
Whose name was blown among strange seas.
Grown old with soft adulteries.
Also her bed was made of green,
Her windows beautiful for glass
That she had made her bed between:
Yea, for pure lust her body was
Made like white summer-coloured grass.
Her raiment was a strong man's spoil;
Upon a table by a bed
She set mine incense and mine oil
To be the beauty of her head
In chambers walled about with red.
Also between the walls she had
Fair faces of strong men portrayed;
All girded round the loins, and clad
With several cloths of woven braid
And garments marvellously made.
Therefore the wrath of God shall be
Set as a watch upon her way;
And whoso findeth by the sea
Blown dust of bones will hardly say
If this were that Aholibah.

LOVE AND SLEEP

Lying asleep between the strokes of night
I saw my love lean over my sad bed,
Pale as the duskiest lily's leaf or head,
Smooth-skinned and dark, with bare throat made to bite,
Too wan for blushing and too warm for white,
But perfect-coloured without white or red.
And her lips opened amorously, and said—
I wist not what, saving one word—Delight.
And all her face was honey to my mouth,
And all her body pasture to mine eyes;
The long lithe arms and hotter hands than fire,
The quivering flanks, hair smelling of the south,
The bright light feet, the splendid supple thighs
And glittering eyelids of my soul's desire.

MADONNA MIA

Under green apple-boughs
That never a storm will rouse,
My lady hath her house
Between two bowers;
In either of the twain
Red roses full of rain;
She hath for bondwomen
All kind of flowers.
She hath no handmaid fair
To draw her curled gold hair
Through rings of gold that bear
Her whole hair's weight;
She hath no maids to stand
Gold-clothed on either hand;
In all the great green land
None is so great.
She hath no more to wear
But one white hood of vair
Drawn over eyes and hair,
Wrought with strange gold,
Made for some great queen's head,
Some fair great queen since dead;
And one strait gown of red
Against the cold.
Beneath her eyelids deep
Love lying seems asleep,
Love, swift to wake, to weep,
To laugh, to gaze;
Her breasts are like white birds,
And all her gracious words
As water-grass to herds
In the June-days.
To her all dews that fall
And rains are musical;
Her flowers are fed from all,
Her joy from these;
In the deep-feathered firs
Their gift of joy is hers,
In the least breath that stirs
Across the trees.

She grows with greenest leaves,
Ripens with reddest sheaves,
Forgets, remembers, grieves,
And is not sad;
The quiet lands and skies
Leave light upon her eyes;
None knows her, weak or wise,
Or tired or glad.
None knows, none understands,
What flowers are like her hands;
Though you should search all lands
Wherein time grows,
What snows are like her feet,
Though his eyes burn with heat
Through gazing on my sweet,
Yet no man knows.
Only this thing is said;
That white and gold and red,
God's three chief words, man's bread
And oil and wine,
Were given her for dowers,
And kingdom of all hours,
And grace of goodly flowers
And various vine.
This is my lady's praise:
God after many days
Wrought her in unknown ways,
In sunset lands;
This was my lady's birth;
God gave her might and mirth
And laid his whole sweet earth
Between her hands.
Under deep apple-boughs
My lady hath her house;
She wears upon her brows
The flower thereof;
All saying but what God saith
To her is as vain breath;
She is more strong than death,
Being strong as love.

THE KING'S DAUGHTER

We were ten maidens in the green corn,
Small red leaves in the mill-water:
Fairer maidens never were born,
Apples of gold for the king's daughter.
We were ten maidens by a well-head,
Small white birds in the mill-water:
Sweeter maidens never were wed,
Rings of red for the king's daughter.
The first to spin, the second to sing,
Seeds of wheat in the mill-water;
The third may was a goodly thing,
White bread and brown for the king's daughter.
The fourth to sew and the fifth to play,
Fair green weed in the mill-water;
The sixth may was a goodly may,
White wine and red for the king's daughter.
The seventh to woo, the eighth to wed,
Fair thin reeds in the mill-water;
The ninth had gold work on her head,
Honey in the comb for the king's daughter.
The ninth had gold work round her hair,
Fallen flowers in the mill-water;
The tenth may was goodly and fair,
Golden gloves for the king's daughter.
We were ten maidens in a field green,
Fallen fruit in the mill-water;
Fairer maidens never have been,
Golden sleeves for the king's daughter.
By there comes the king's young son,
A little wind in the mill-water;
"Out of ten maidens ye'll grant me one,"
A crown of red for the king's daughter.
"Out of ten mays ye'll give me the best,"
A little rain in the mill-water;
A bed of yellow straw for all the rest,
A bed of gold for the king's daughter.
He's ta'en out the goodliest,
Rain that rains in the mill-water;
A comb of yellow shell for all the rest,

A comb of gold for the king's daughter.
He's made her bed to the goodliest,
Wind and hail in the mill-water;
A grass girdle for all the rest,
A girdle of arms for the king's daughter.
He's set his heart to the goodliest,
Snow that snows in the mill-water;
Nine little kisses for all the rest,
An hundredfold for the king's daughter.
He's ta'en his leave at the goodliest,
Broken boats in the mill-water;
Golden gifts for all the rest,
Sorrow of heart for the king's daughter.
"Ye'll make a grave for my fair body,"
Running rain in the mill-water;
"And ye'll streek my brother at the side of me,"
The pains of hell for the king's daughter.

AFTER DEATH

The four boards of the coffin lid
Heard all the dead man did.
The first curse was in his mouth,
Made of grave's mould and deadly drouth.
The next curse was in his head,
Made of God's work discomfited.
The next curse was in his hands,
Made out of two grave-bands.
The next curse was in his feet,
Made out of a grave-sheet.
"I had fair coins red and white,
And my name was as great light;
I had fair clothes green and red,
And strong gold bound round my head.
But no meat comes in my mouth,
Now I fare as the worm doth;
And no gold binds in my hair,
Now I fare as the blind fare.
My live thews were of great strength,
Now am I waxen a span's length;
My live sides were full of lust,
Now are they dried with dust."
The first board spake and said:
"Is it best eating flesh or bread?"
The second answered it:
"Is wine or honey the more sweet?"
The third board spake and said:
"Is red gold worth a girl's gold head?"
The fourth made answer thus:
"All these things are as one with us."
The dead man asked of them:
"Is the green land stained brown with flame?
Have they hewn my son for beasts to eat,
And my wife's body for beasts' meat?
Have they boiled my maid in a brass pan,
And built a gallows to hang my man?"
The boards said to him:
"This is a lewd thing that ye deem.
Your wife has gotten a golden bed,
All the sheets are sewn with red.

Your son has gotten a coat of silk,
The sleeves are soft as curded milk.
Your maid has gotten a kirtle new,
All the skirt has braids of blue.
Your man has gotten both ring and glove,
Wrought well for eyes to love."
The dead man answered thus:
"What good gift shall God give us?"
The boards answered him anon:
"Flesh to feed hell's worm upon."

MAY JANET
(BRETON)

"Stand up, stand up, thou May Janet,
And go to the wars with me."
He's drawn her by both hands
With her face against the sea.
"He that strews red shall gather white,
He that sows white reap red,
Before your face and my daughter's
Meet in a marriage-bed.
"Gold coin shall grow in the yellow field,
Green corn in the green sea-water,
And red fruit grow of the rose's red,
Ere your fruit grow in her."
"But I shall have her by land," he said,
"Or I shall have her by sea,
Or I shall have her by strong treason
And no grace go with me."
Her father's drawn her by both hands,
He's rent her gown from her,
He's ta'en the smock round her body,
Cast in the sea-water.
The captain's drawn her by both sides
Out of the fair green sea;
"Stand up, stand up, thou May Janet,
And come to the war with me."
The first town they came to
There was a blue bride-chamber;
He clothed her on with silk
And belted her with amber.
The second town they came to
The bridesmen feasted knee to knee;
He clothed her on with silver,
A stately thing to see.
The third town they came to
The bridesmaids all had gowns of gold;
He clothed her on with purple,
A rich thing to behold.
The last town they came to
He clothed her white and red,

With a green flag either side of her
And a gold flag overhead.

THE BLOODY SON
(FINNISH)

"O where have ye been the morn sae late,
My merry son, come tell me hither?
O where have ye been the morn sae late?
And I wot I hae not anither."
"By the water-gate, by the water-gate,
O dear mither."
"And whatten kin' o' wark had ye there to make,
My merry son, come tell me hither?
And whatten kin' o' wark had ye there to make?
And I wot I hae not anither."
"I watered my steeds with water frae the lake,
O dear mither."
"Why is your coat sae fouled the day,
My merry son, come tell me hither?
Why is your coat sae fouled the day?
And I wot I hae not anither."
"The steeds were stamping sair by the weary banks of clay,
O dear mither."
"And where gat ye thae sleeves of red,
My merry son, come tell me hither?
And where gat ye thae sleeves of red?
And I wot I hae not anither."
"I have slain my ae brither by the weary waterhead,
O dear mither."
"And where will ye gang to mak your mend,
My merry son, come tell me hither?
And where will ye gang to mak your mend?
And I wot I hae not anither."
"The warldis way, to the warldis end,
O dear mither."
"And what will ye leave your father dear,
My merry son, come tell me hither?
And what will ye leave your father dear?
And I wot I hae not anither."

"The wood to fell and the logs to bear,
For he'll never see my body mair,
O dear mither."
"And what will ye leave your mither dear,
My merry son, come tell me hither?
And what will ye leave your mither dear?
And I wot I hae not anither."
"The wool to card and the wool to wear,
For ye'll never see my body mair,
O dear mither."
"And what will ye leave for your wife to take,
My merry son, come tell me hither?
And what will ye leave for your wife to take?
And I wot I hae not anither."
"A goodly gown and a fair new make,
For she'll do nae mair for my body's sake,
O dear mither."
"And what will ye leave your young son fair,
My merry son, come tell me hither?
And what will ye leave your young son fair?
And I wot ye hae not anither."
"A twiggen school-rod for his body to bear,
Though it garred him greet he'll get nae mair,
O dear mither."
"And what will ye leave your little daughter sweet,
My merry son, come tell me hither?
And what will ye leave your little daughter sweet?
And I wot ye hae not anither."
"Wild mulberries for her mouth to eat,
She'll get nae mair though it garred her greet,
O dear mither."
"And when will ye come back frae roamin',
My merry son, come tell me hither?
And when will ye come back frae roamin'?
And I wot I hae not anither."
"When the sunrise out of the north is comen,
O dear mither."
"When shall the sunrise on the north side be,
My merry son, come tell me hither?
When shall the sunrise on the north side be?
And I wot I hae not anither."

"When chuckie-stanes shall swim in the sea,
O dear mither."
"When shall stanes in the sea swim,
My merry son, come tell me hither?
When shall stanes in the sea swim?
And I wot I hae not anither."
"When birdies' feathers are as lead therein,
O dear mither."
"When shall feathers be as lead,
My merry son, come tell me hither?
When shall feathers be as lead?
And I wot I hae not anither."
"When God shall judge between the quick and dead,
O dear mither."

THE SEA-SWALLOWS

This fell when Christmas lights were done,
(Red rose leaves will never make wine)
But before the Easter lights begun;
The ways are sair fra' the Till to the Tyne.
Two lovers sat where the rowan blows
And all the grass is heavy and fine,
By the gathering-place of the sea-swallows
When the wind brings them over Tyne.
Blossom of broom will never make bread,
Red rose leaves will never make wine;
Between her brows she is grown red,
That was full white in the fields by Tyne.
"O what is this thing ye have on,
Show me now, sweet daughter of mine?"
"O father, this is my little son
That I found hid in the sides of Tyne.
"O what will ye give my son to eat,
Red rose leaves will never make wine?"
"Fen-water and adder's meat."
The ways are sair fra' the Till to the Tyne.
"Or what will ye get my son to wear?"
(Red rose leaves will never make wine.)
"A weed and a web of nettle's hair."

The ways are sair fra' the Till to the Tyne.
"Or what will ye take to line his bed?"
(Red rose leaves will never make wine.)
"Two black stones at the kirkwall's head."
The ways are sair fra' the Till to the Tyne.
"Or what will ye give my son for land?"
(Red rose leaves will never make wine.)
"Three girl's paces of red sand."
The ways are sair fra' the Till to the Tyne.
"Or what will ye give me for my son?"
(Red rose leaves will never make wine.)
"Six times to kiss his young mouth on."
The ways are sair fra' the Till to the Tyne.
"But what have ye done with the bearing-bread,
And what have ye made of the washing-wine?
Or where have ye made your bearing-bed,
To bear a son in the sides of Tyne?"
"The bearing-bread is soft and new,
There is no soil in the straining wine;
The bed was made between green and blue,
It stands full soft by the sides of Tyne.
"The fair grass was my bearing-bread,
The well-water my washing-wine;
The low leaves were my bearing-bed,
And that was best in the sides of Tyne."
"O daughter, if ye have done this thing,
I wot the greater grief is mine;
This was a bitter child-bearing,
When ye were got by the sides of Tyne.
"About the time of sea-swallows
That fly full thick by six and nine,
Ye'll have my body out of the house,
To bury me by the sides of Tyne.
"Set nine stones by the wall for twain,"
(Red rose leaves will never make wine)
"For the bed I take will measure ten."
The ways are sair fra' the Till to the Tyne.
"Tread twelve girl's paces out for three,"
(Red rose leaves will never make wine)
"For the pit I made has taken me."
The ways are sair fra' the Till to the Tyne.

THE YEAR OF LOVE

There were four loves that one by one,
Following the seasons and the sun,
Passed over without tears, and fell
Away without farewell.
The first was made of gold and tears,
The next of aspen-leaves and fears,
The third of rose-boughs and rose-roots,
The last love of strange fruits.
These were the four loves faded. Hold
Some minutes fast the time of gold
When our lips each way clung and clove
To a face full of love.
The tears inside our eyelids met,
Wrung forth with kissing, and wept wet
The faces cleaving each to each
Where the blood served for speech.
The second, with low patient brows
Bound under aspen-coloured boughs
And eyes made strong and grave with sleep
And yet too weak to weep—
The third, with eager mouth at ease
Fed from late autumn honey, lees
Of scarce gold left in latter cells
With scattered flower-smells—
Hair sprinkled over with spoilt sweet
Of ruined roses, wrists and feet
Slight-swathed, as grassy-girdled sheaves
Hold in stray poppy-leaves—
The fourth, with lips whereon has bled
Some great pale fruit's slow colour, shed
From the rank bitter husk whence drips
Faint blood between her lips—
Made of the heat of whole great Junes
Burning the blue dark round their moons
(Each like a mown red marigold)
So hard the flame keeps hold—
These are burnt thoroughly away.
Only the first holds out a day
Beyond these latter loves that were

Made of mere heat and air.
And now the time is winterly
The first love fades too: none will see,
When April warms the world anew,
The place wherein love grew.

DEDICATION
1865

The sea gives her shells to the shingle,
The earth gives her streams to the sea:
They are many, but my gift is single,
My verses, the firstfruits of me.
Let the wind take the green and the grey leaf,
Cast forth without fruit upon air;
Take rose-leaf and vine-leaf and bay-leaf
Blown loose from the hair.
The night shakes them round me in legions,
Dawn drives them before her like dreams;
Time sheds them like snows on strange regions,
Swept shoreward on infinite streams;
Leaves pallid and sombre and ruddy,
Dead fruits of the fugitive years;
Some stained as with wine and made bloody,
And some as with tears.
Some scattered in seven years' traces,
As they fell from the boy that was then;
Long left among idle green places,
Or gathered but now among men;
On seas full of wonder and peril,
Blown white round the capes of the north;
Or in islands where myrtles are sterile
And loves bring not forth.
O daughters of dreams and of stories
That life is not wearied of yet,
Faustine, Fragoletta, Dolores,
Félise and Yolande and Juliette,
Shall I find you not still, shall I miss you,
When sleep, that is true or that seems,
Comes back to me hopeless to kiss you,
O daughters of dreams?
They are past as a slumber that passes,
As the dew of a dawn of old time;
More frail than the shadows on glasses,
More fleet than a wave or a rhyme.
As the waves after ebb drawing seaward,
When their hollows are full of the night,

So the birds that flew singing to me-ward
Recede out of sight.
The songs of dead seasons, that wander
On wings of articulate words;
Lost leaves that the shore-wind may squander,
Light flocks of untameable birds;
Some sang to me dreaming in class-time
And truant in hand as in tongue;
For the youngest were born of boy's pastime,
The eldest are young.
Is there shelter while life in them lingers,
Is there hearing for songs that recede,
Tunes touched from a harp with man's fingers
Or blown with boy's mouth in a reed?
Is there place in the land of your labour,
Is there room in your world of delight,
Where change has not sorrow for neighbour
And day has not night?
In their wings though the sea-wind yet quivers,
Will you spare not a space for them there
Made green with the running of rivers
And gracious with temperate air;
In the fields and the turreted cities,
That cover from sunshine and rain
Fair passions and bountiful pities
And loves without stain?
In a land of clear colours and stories,
In a region of shadowless hours,
Where earth has a garment of glories
And a murmur of musical flowers;
In woods where the spring half uncovers
The flush of her amorous face,
By the waters that listen for lovers,
For these is there place?
For the song-birds of sorrow, that muffle
Their music as clouds do their fire:
For the storm-birds of passion, that ruffle
Wild wings in a wind of desire;
In the stream of the storm as it settles
Blown seaward, borne far from the sun,
Shaken loose on the darkness like petals

Dropt one after one?
Though the world of your hands be more gracious
And lovelier in lordship of things
Clothed round by sweet art with the spacious
Warm heaven of her imminent wings,
Let them enter, unfledged and nigh fainting,
For the love of old loves and lost times;
And receive in your palace of painting
This revel of rhymes.
Though the seasons of man full of losses
Make empty the years full of youth,
If but one thing be constant in crosses,
Change lays not her hand upon truth;
Hopes die, and their tombs are for token
That the grief as the joy of them ends
Ere time that breaks all men has broken
The faith between friends.
Though the many lights dwindle to one light,
There is help if the heaven has one;
Though the skies be discrowned of the sunlight
And the earth dispossessed of the sun,
They have moonlight and sleep for repayment,
When, refreshed as a bride and set free,
With stars and sea-winds in her raiment,
Night sinks on the sea.